Camila de Oliveira Barros
Kátia E. de S. Miranda
Wagna P. C. dos Santos

Development of cereal bars based on legumes

AF144432

Camila de Oliveira Barros
Kátia E. de S. Miranda
Wagna P. C. dos Santos

Development of cereal bars based on legumes

Family farming in food production and valuing local culture and eating habits

ScienciaScripts

Cover image: www.ingimage.com

This book is a translation from the original published under ISBN 978-620-2-04902-3.

Publisher:
Sciencia Scripts
is a trademark of
Dodo Books Indian Ocean Ltd. and OmniScriptum S.R.L publishing group

120 High Road, East Finchley, London, N2 9ED, United Kingdom
Str. Armeneasca 28/1, office 1, Chisinau MD-2012, Republic of Moldova, Europe
Printed at: see last page
ISBN: 978-620-7-24013-5

SUMMARY

1 INTRODUCTION

Cereal bars are considered practical and provide health benefits. They were introduced to the Brazilian market in the 90s. Depending on their composition, they can be a source of vitamins, minerals, proteins, complex carbohydrates and high fiber content. The appreciation of healthier eating habits and the search for a better quality of life by consumers have led to cereal bars acquiring a large space in the food market as they replace other products with less nutritional value (MARQUES, 2013), because they are easy to consume since they don't require any additional preparation and because they are sold in individual packets (SREBERNICH, 2016).

According to Sousa (2014), the regions with the highest consumption of cereal bars in Brazil are the South, with a preference for the chocolate version, and the Northeast, where those with fruit and cereals in the composition predominate, respectively 24.5% and 18% in terms of sales. In Brazil, cereal bars were initially aimed at fans of extreme sports and, over time, they have won over a variety of audiences, such as women, children, the elderly and weekend athletes (FREITAS; MORETTI, 2006, SOUSA, 2014).

Due to the wide variety of products and the search to satisfy consumers without damaging the most appreciated sensory attributes, new alternatives to improve the nutritional quality of cereal bars have been developed using new food ingredients (SANTOS, 2010). In this sense, legumes may represent considerable potential as a raw material to be used in the production of these products.

There is a wide variety of legumes, especially in terms of the shape, size and color of the grains, and in the Brazilian market this difference is very evident. Legume grains are usually recognized and identified as "beans". A bean is said to be of quality when it is judged on three technological points: commercial, culinary and nutritional (CHAVES; BASSINELLO, 2014).

Present on the plates of Brazilians, even after a drop in consumption over the last 40 years due to an increase in the consumption of industrialized products, according to the 2008-2009 Household Budget Survey (POF), the search for alternatives more

2

suited to consumer demands has led to new bean products being developed, adding value to the processed grain, thus offering consumers greater practicality of consumption and semi-ready products (MARQUEZI, 2013).

In this way, the use of mangrove, caupi and andu bean flours in the development of cereal bars contributes to the combination of various ingredients with specific functionality, making them nutritious and functional and valuing local culture and eating habits, adding value to regional foods, and reducing post-harvest losses for small farmers.

Thus, the development of cereal bars from these flours, which are easy to process and reproduce for small farmers, serves as a basis for the creation of new products, undergoing changes according to the eating habits of the consumers for whom they are intended.

2 THE IMPORTANCE OF LEGUMES IN FAMILY FARMING

2.1 FAMILY FARMING

Family farming encompasses great cultural, social and economic diversity and can range from the traditional peasantry to modernized small-scale production. Known as small producers, small farmers, settlers, peasants, quilombolas, land reform settlers, traditional peoples and communities, among others, their definition is linked to the number of employees and the size of the property (CRUZ et al., 2006).

According to the Food and Agriculture Organization of the United Nations - FAO (2014), family farming consists of a means of organizing agricultural, forestry, fishing, pastoral and aquaculture production that is managed and operated by a family and predominantly dependent on family labour, both women and men.

According to Law No. 11.326 of July 24, 2006, the characteristics of a family farmer are as follows:

[...] farmers who do not have an area larger than 4 (four) fiscal modules, who predominantly use their own family's labor force in economic activities; have a minimum percentage of their family's income originating from economic activities, as defined by the Executive Branch and run their establishment or enterprise with their family (BRASIL, 2006).

In Brazil, the sector encompasses 4.3 million production units (84% of rural establishments) and 14 million employed people, which represents around 74% of all occupations distributed over 80,250,453 hectares (25% of the total area) (EMBRAPA, 2014). The Northeast concentrates the largest number of family farmers, representing 50.1% of the national total (SILVA; COSTA, 2012).

This sector is important in terms of job absorption and food production, responsible for around 70% of the food consumed throughout Brazil (MDA, 2015), currently supplying the Brazilian market with: cassava (87%), beans (70%), pork (59%), milk (58%), poultry (50%) and corn (46%) (PORTAL BRASIL, 2015), as well as being a factor in reducing the rural exodus and a source of resources for families with lower incomes, it also makes a significant contribution to generating wealth in the country (GUILHOTO et al., 2007), as well as including new social and environmental

4

functions, and even the preservation of the landscape and cultural traditions (MENDES et al., 2005). Family farming is also responsible for a portion of the foodstuffs destined for school meals thanks to the incentive provided by Law No. 11.947/2009, allowing healthy food with a regional link to be consumed daily by public school students throughout Brazil (FNDE, 2017).

In this context, in the state of Bahia, the municipality of Cruz das Almas is part of the Recôncavo Territory Family Farming Cooperative - COOAFATRE, which is also made up of the municipalities of Sao Félix, Sao Felipe and Maragojipe, comprising the Recôncavo Baiano Territory (SILVA; COSTA, 2012).

The crops produced are cassava (*Manihot esculenta* C.), yams (*Dioscorea cayennensis* Lam.), corn (*Zea mays* L.), peanuts (*Arachis hypogaea L),* beans (*Phaseolus vulgaris* L.), breadfruit (*Artocarpus altilis*), vegetables, purple potatoes, sweet potatoes (*Ipomoea batatas* Lam.), andu, mangalô (SILVA; COSTA, 2012).

The Brazilian Agricultural Research Corporation (Embrapa) also points to the need to implement technologies that ensure sustainable and competitive agricultural production, making producers more competitive in a globalized market.

2.2 BEANS (Family *Fabaceae*)

In 2016, the United Nations declared the International Year of Pulses in recognition of their fundamental role as a source of income for millions of family farmers, in food and nutritional security, in adapting to climate change by fixing nitrogen in the soil, in human health and as a key to tackling problems such as obesity and hunger (FAO, 2017).

According to Salvador (2015), Brazil is the world's 3rd largest producer of beans, accounting for 11% of production, behind Myanmar with 13% and India with 14%. As for the countries that make up Mercosur, Brazil ranks first as the largest producer and consumer with around 3.1 million tons per year (CONAB, 2015).

When consumed together with cereals, legumes form a complete protein, which is cheaper than animal protein and therefore more accessible to families with low economic resources (FAO, 2017). It is a perfect combination with rice, as both provide the amino acids (lysine and methionine, 3:1) that help form the proteins in the human body, and, being one of the components of the Brazilian food basket, national production of beans is heavily geared towards domestic consumption (IBGE, 2011). The reduction in consumption of this product is due to the process of urbanization, changes in eating habits, and the greater demand for quick-preparation products, since this legume, after a certain storage time, requires more cooking time, which causes consumers to reject it (IBGE, 2011; RUAS, 2015).

The culinary characteristics of a bean that are desirable to consumers are related to rapid hydration, low cooking time, the production of a thick broth, good flavor and texture, moderately cracked grains, a thin shell and good color stability (CHAVES; BASSINELLO, 2014).

According to Natabirwa, Katende and Lungaho (2014):

Beans are rich sources of vital nutrients including high protein (18-30%) and soluble fiber, which is important for improving the movement of food in the intestine and controlling diabetes. In addition, they provide iron, zinc, folic acid, magnesium, manganese and B vitamins. However, beans are often consumed in just one way by almost everyone, especially at home. When this consumption habit is combined with unlimited forms of preparation, the quantities consumed rarely meet nutritional requirements (NATABIRWA, KATENDE AND LUNGAHO, 2014, p.(i)).

Studies show that due to their high concentration of nutrients, beans are health-promoting and reduce the risk of developing certain diseases, such as heart disease, obesity and many types of cancer.

The dynamics of bean production involve three harvests: the wet season, with harvests concentrated in the months of December to March, grown mainly in the South and Southeast and in the Irecè region of Bahia; the dry season or "safrinha" in the months of April to July, grown in the northeast region, and the winter season, which is offered to the market in the months of July to October, when irrigated beans are predominantly grown in the states of Minas Gerais, Sao Paulo, Espirito Santo, Goiàs/Federal District and western Bahia (FERREIRA; PELOSO; FARIA, 2003).

Two species considered to be socially and economically important by the Ministry of

6

Agriculture, Livestock and Food Supply are cultivated in Brazil: *Phaseolus vulgaris,* known as the common bean, and *Vigna unguiculata*, known as the green bean (MAPA, 2008). According to Freire Filho and Rocha (2016), green beans correspond to pods around maturity, i.e. just before or just after the stage at which they stop accumulating photosynthates and begin the process of natural dehydration. The difference between the species will be associated with the physiological ripeness indicators of the seeds.

According to Marquezi (2013), there are few studies that relate the different traditional applications or even products based on beans to the characteristics of the raw material, which is fundamental for the development of new products, placing beans in a prominent position in proportion to their important nutritional composition.

Marquezi (2013) states that bean flours have technological characteristics such as neutral pH, foam formation, emulsifying capacity and emulsion stability, and suggests the use of these flours in the development of new products.

2.2.1 Bitter Mangaloo Beans *(Lablab purpureus (L.)* Sweet*)*

The bitter mangrove bean (*Lablab purpurlus* (L.) Sweet) (Figures 1 and 2), also known as orelha-de-padre, feijao-de-pedra and lablab. Originally from Africa, it is grown mainly in the Northeast region and is a legume with multiple uses, whether for human food, as fodder for animal feed, or for inclusion in conservation farming systems as a green manure or cover crop, and is commonly intercropped with maize (BRASIL, 2015).

Figura 1: Bitter mangrove bean (Lablab purpureus (L.) Sweet) on broad beans.

Source: Authors

Figura 2: Frozen threshed bitter mangrove beans *(Lablab purpureus* (L.) Sweet).

Source: Authors

Currently, the bitter mangrove bean (*Lablab purpurlus* (*L.*) Sweet) is on the list of unconventional vegetables, i.e. vegetables that at one time were widely consumed by the population but, due to changes in eating behavior, have become less economically and socially significant, losing space and market share to other vegetables (EPAMIG, 2016).

In cooking, the pods and ripened grains are consumed. They can be used to enhance salads, soups and stews, but because they have a slight bitterness, the grains should be blanched before being cooked (BRASIL, 2015).

According to Rubatzky and Yamaguchi (1997), the nutritional composition of *lablab* bean seeds *in natura* contains 100 g of edible part: water 87 g, calories 193 kJ (46 kcal), proteins 2.9 g, lipids 0.45 g, carbohydrates 2.9 g, fiber 1.5 g, Ca 0.6 mg, Mg 37 mg, P 59mg, Fe 1.2 mg, vitamin A 210 mg, thiamine 0.9 mg, riboflavin 0.08 mg, niacin 0.6 mg and ascorbic acid 11 mg.

8

2.2.2 Cowpea (*Vigna unguiculatra* (L.) Walp)

The caupi bean (*Vigna unguiculata* (L.) Walp) (Figures 3 and 4), also known as feijao-de-corda, feijao-verde, feijao-caupi, caupi, feijao-macàçar (macassar), feijao-fradinho, fradinho e vigna, trepa-pau, feijao gurutuba, feijao catador, feijao-de-praia, stands out according to Lima et al, (2004) as one of the main crops in the northeast and north of the country, having been introduced to Brazil by the Spanish and slaves.

Cultivated in tropical Africa, South America and Asia, this type of bean is the staple food of many rural populations due to its high nutritional value, in terms of protein and energy, and its easy adaptation to low-fertility soils and periods of prolonged drought. In Bahia, it is widely used in the preparation of acarajè, a typical food of the state (BRASIL, 2015).

According to Freitas (2011) the main producing states in the Northeast are: Cearà, Bahia, Piaui, Pernambuco, Paraiba, Rio Grande do Norte and Maranhao and in the North: Amapà, Parà, Rondônia and Roraima, whose production is destined for domestic consumption and is of great importance as food and in generating income for family farming.

Figure 3: Caupi bean (*Vigna unguiculatra* (L.) Walp) on broad beans.

Source: Authors

Figura 4: Frozen Caupi beans (*Vigna unguiculatra* (L.) Walp).

9

According to Embrapa (2002) the cowpea (*Vigna unguiculata (L.)*) is an excellent source of protein (23%-25% on average) and has all the essential amino acids, carbohydrates (62% on average), vitamins and minerals, as well as being high in dietary fiber, low in fat (2% lipid content on average) and containing no cholesterol (Table 1).

Table 1: Agronomic food characteristics of cowpea (*Vigna unguiculata (L.)*).

		In 100g (TACO)
Proteins	23% - 25%	20,2 g
Carbohydrates	62%	61,2 g
Fats	2%	2,4 g
Cholesterol	0%	NA
Calories	323 - 339 kcal/100g (TACO, 2006, FROTA et al., 2008)	
Glycemic index	Low, 36/100 Glucose	-

Source: Modified from GÓES, CAVALCANTE, 2013.

With a regional market, the commercialization of caupi is restricted to dried grains, green grains (hydrated) and seeds. There are already some initiatives for the industrial processing of caupi to produce flour and pre-cooked and frozen products (RIBEIRO, 2002). Cashew bean flour is used in enriched foods such as cookies and rocamboles because it has good acceptability and stability and a high protein content (FROTA et al., 2010).

2.2.3 Andu Beans (*Cajanus cajan* (L) Huth)

The andu bean (*Cajanus cajan* (L) Huth), Figures 5 and 6, has many uses and is found mainly in the backyards of many country towns. It has a high protein content and significant levels of calcium, iron, magnesium and phosphorus (Table 2) (AZEVEDO; RIBEIRO; AZEVEDO, 2007). Known as feijao-andu, andu, guando, guandeiro, cuandu, feijao-cuandu, feijào-de-àrvore, ervilha-de- angola, ervilha-de sete-anos, ervilha-do-congo. It was introduced to Brazil and the Guianas on the slave routes from Africa (BRASIL, 2015).

Figura 5: Andu beans (*Cajanus cajan (*L) Huth) in fava beans.

Source: Authors

Figura 6: Andu beans (*Cajanus cajan (*L) Huth) threshed and frozen.

Source: Authors

Because it has green, very palatable grains, it has been used as a substitute for peas and is prepared with meat, farofas or stir-fries. They can also be preserved in brine or frozen (BRASIL, 2015). Due to its functional properties of protein solubility as a function of pH, water and oil absorption capacity, gel formation capacity and emulsion formation and stability, it is recommended for use in bakery and confectionery products (MIZUBUTI, et al., 2000).

11

There are several applications for the Andu bean crop. According to Azevedo, Ribeiro and Azevedo (2007):

[...] it can be used for the most diverse purposes: as a soil improvement plant, in the recovery of degraded areas, as a phytoremediation plant, in the renovation of pastures, in the feeding of domestic animals and livestock, also widely used in human food (AZEVEDO, RIBEIRO E AZEVEDO, 2007, p. 82).

Table 2: Nutritional analysis in 100 g of pigeonpea (*Cajanus cajan* (L) Huth)

Energy	Proteins	Lipids	Carbohydrates	Fibers	Calcium	Phosphorus
344(Kcal)	19(g)	2,1(g)	64(g)	21,3(g)	3.5(mg)	269(mg)
Iron	Retinol	Vit. B1	Vit. B2	Niacin	Vit. C	
1.9 (mg)	NA	1.06 (mg)	Tr	2.7 (mg)	1.5 (mg)	

Source: TACO (2011); Brazil (2015)

The Food Guide for the Brazilian Population (Brazil, 2014) emphasizes that the alternation between different types of beans and other legumes amplifies the supply of nutrients, bringing new flavors and diversity to the diet and, by having a high fiber content and a moderate amount of calories per gram, gives these foods a high satiety power, thus avoiding excess food consumption. In addition, products made from beans represent an alternative for people on special diets (gluten-free, vegetarian, among others), providing a variety of nutrients.

3 CEREAL BARS

3.1 CEREAL BAR: A TREND?

Cereal bars are consumed almost six times more than eight years *ago* (DEGÀSPARI; BLINDER; MOTTIN, 2008). Classified as "snacks", they are defined as small, light or substantial meals (SAMPAIO, 2009). Depending on their composition, in terms of caloric values, cereal bars are not recommended as a substitute for main meals, but should be eaten as a snack, afternoon snack or supper.

There are several types of cereal bars on the market: conventional (fibrous bars); meal replacements, created especially for those who want to lose weight, their formulation aims to maintain a complete nutritional balance, being replacements for the morning or afternoon snack; energy and protein bars recommended particularly for sportsmen and athletes; diets (sugar-free) and light, with a reduction of at least 25% in some specific nutrient and finally cereal bars with seeds, rich in mono- and polyunsaturated fatty acids (LOUIZE, 2016).

There are various definitions of cereal bars, according to Sampaio (2009), Guimaraes e Silva (2009), Gutkoski et al. (2007) among others, they are foods made by compacting or extruding cereal dough or a mixture of dry ingredients (cereal or cookie, cornflakes, rice flakes, oats) with a binding agent (or binding syrup), containing dried fruit (dehydrated), with or without nuts, with or without chocolate coating and flavors that give the final product distinct technological characteristics. It is a particular category of confectionery products, usually rectangular in shape, sold in individual units for consumption by a single person.

Cereal bars are multi-component and can be very complex in their formulation. All the ingredients that make it up are combined to guarantee flavor, texture and characteristic physical properties (GUTKOSKI et al., 2007).

According to studies by Degâspari; Blinder; Mottin (2008), the biggest consumers of cereal bars are women and the age of consumers of both sexes is less than 44 years. The studies also show that this is a relatively high-priced product, which is consumed less by people on lower incomes and can be considered an elite product.

3.2 BASIC INGREDIENTS

There is a wide variety of ingredients that can be used to make cereal bars, seeking to relate the product to health benefits such as: textured protein, wheat germ and oats, supplemented with vitamin C and E, containing residues from the manufacture of cassava flour and yellow passion fruit, presenting specific and/or functional functions, being altered depending on the composition of each one and the flavor (SANTOS, 2010).

- **Cereals**

Cereals are edible seeds or grains from the grass family, *Gramineae,* such as wheat, rice, rye and oats. They are staple foods and perform important functions, being sources of energy, carbohydrates, protein (6-15%), fiber, vitamin E, B vitamins, magnesium, zinc and bioactive substances for developed and developing countries (BRIGID MCKEVITH, 2004).

Present in most cereal bars, rice flakes are by-products of the polishing of brown rice using the thermoplastic extrusion technique with or without the addition of other ingredients. Extrusion results in the gelatinization of starch, denaturation of proteins and the formation of complexes between starches, lipids and proteins (TRAMUJAS, 2015). Rice flakes are crunchy and have functional effects, making them beneficial for use in food products due to their antioxidant effect, neutralizing the release of free radicals during intense exercise and helping to release endorphins, which give a feeling of well-being (GUTKOSKI; TROMBETTA, 1999).

Multifunctional, oats (*Avena sativa* L) are an excellent source of protein (12 to 14%), lipids (essential Iinoleic acid), antioxidants (tocopherol, phenolic acids and derivatives), B vitamins, calcium, iron, with a high content of dietary fiber and β-glucan (AHMAD et al., 2014). As one of the main ingredients in cereal bars (SAMPAIO, 2009), oats contribute to stability, flavor, increase the fiber content of food products, as well as other functional and bioactive properties (AHMAD et al.,

2014).

Oats are most commonly marketed in the form of flakes (TRAMUJAS, 2015). β-glucans, present in oats (*Avena sativa L*), have properties related to viscosity, such as increasing the viscosity of intestinal fluids, combined with insulin can replace fats, in the food industry is being widely considered with the dual purpose of increasing the fiber content of food products and enhancing their health properties (AHMAD et al., 2014).

- **Flours**

According to Resolution RDC No. 263 of September 22, 2005, flours are products obtained from edible parts of one or more species of cereals, legumes, fruits, seeds, tubers and rhizomes by milling and/or other technological processes considered safe for food production.

The moisture content of flour directly influences its quality and that of the final product. The maximum grain moisture allowed in Brazil is 13% (BRASIL, 2001), and according to Brazilian legislation, the maximum moisture limit for wheat flour is 15% (BRASIL, 2005). The development of legume-based flours can provide nutritional enrichment for foods traditionally available on the market.

- **Banana raisins**

In cereal bars, the use of dried or desiccated fruit helps to improve the soluble and insoluble fiber profile of the product and improve its technological and functional properties (GUIMARÂES; SILVA, 2009; MUNHOZ, 2013).

Fruit drying, or raisin production, is a practice used to make use of surplus production which, as well as adding value to the product, extends its useful life and can be stored and marketed outside the harvest season. It is obtained by partially losing the water from the ripe fruit, whole or in pieces, using appropriate technological processes (PIOVESANA, 2011).

In cereal bars, raisined bananas are used to enhance the flavor, increase the fiber content and modify the energy content (GUIMARÂES; SILVA, 2009).

- **Sugars**

Sugar is a staple of Brazilian culture. There are different types and ways of consuming it, whether added to food or culinary preparations. Crystal sugar is presented in the form of large, transparent crystals, with a slight refining process (OETTERER; SARMENTO, 2006). When making syrup, which is responsible for the agglomeration of solid ingredients and also for the sweet taste, using only sucrose can result in a dry, hard and grainy product, due to its solubility limit of around 67% (GALLI et al., 1996).

Inverted sugar gets its name from the inversion of the optical power of the solution with the addition of acid, an older and more economical method. Acid (the catalyst for the reaction) causes the glycosidic bond of sucrose to break, forming glucose and fructose. A sucrose solution rotates the plane-polarized light to the right (positive direction) and, as the sucrose is hydrolyzed by acid or enzymes into reducing sugars, the plane-polarized light rotates to the left (negative direction) (PODADERA, 2007).

The acid used to produce invert sugar is citric acid from lemon juice and other fruits, vinegar and cream of tartar, whose action is accelerated by boiling (PHILIPPI, 2014).

Invert sugar is a natural sugar that has a sweetening capacity around 70% higher than sucrose, has antioxidant properties, is more resistant to microbiological contamination, has high hygroscopicity and lower viscosity, is resistant to crystallization and can be stored at high concentrations (80%). This sugar eliminates the need for pasteurizing, dissolving and filtering the sugar, and stimulates the Maillard reaction (ALMEIDA, 2003).

Maltodextrin is defined by the *Food and Drug Administration* (FDA) in the USA as an unsweetened, nutritious saccharide polymer consisting of D-glucose units linked primarily by $\alpha(1\text{-}4)$ bonds and having a dextrose equivalent (DE) of less than 20. It is prepared as a fine white powder or concentrated solution by partial hydrolysis of corn starch, potato starch or rice starch with safe and suitable acids and enzymes.

In addition to being a thickening agent, maltodextrin is also used in the food industry to help with spray-drying, as a fat substitute, as a film former, to control freezing, to prevent crystallization and as a nutritional supplement used as an ergogenic resource for physical activity practitioners (COUTINHO, 2007).

The combination of these sugars in cereal bars is responsible for the binding of the cereals, their moistness and flavor (MAESTRI; FERREIRA; PASQUALLI; 2012).

- **Soy lecithin**

Soy lecithin is a phospholipid used as a natural emulsifier, stabilizing agent, emollient and as a form of stable foams in food production (AMARAL, PEALEZ, LIMA, 2011). Due to its chemical structure made up of a mixture of 21% phosphatidylcholine, 22% loslatidylethanolamine, 19% loslatidyl inositol, combined with other substances such as triglycerides, fatty acids and carbohydrates (FRAGON, 2016), it can be solubilized in polar and apolar solutions, which creates great versatility in the use of this ingredient.

In cereal bars, soy lecithin acts as a binding agent, helping to mix and interact the components of the flour and other ingredients, improving volume and texture (FOOD INGREDIENTS BRASIL, 2013).

- **Sodium Chloride**

The Brazilian Society of Hypertension and the World Health Organization (Gowdak, 2018) recommend a daily sodium intake of 2000 mg of sodium/day, which is equivalent to 5 grams of sodium chloride/day. Sodium chloride is used in the food industry as a preservative and flavor enhancer (TRAMUJAS, 2015). It is used in cereal bars to add flavor.

- **Soybean oil**

Soybean oil, known as cooking or salad oil, is light in color and mild in taste. It contains large quantities of polyunsaturated acids, added natural and synthetic antioxidants and pigments, is used as an emulsifier and has stabilizing properties due to its compounds (HAMMOND et al., 2005).

17

It is added to cereal bars for softness and shine, protecting the cereals from moisture by forming a film on the surface (MAESTRI; FERREIRA; PASQUALLI; 2012).

4 BAR PRODUCTION PROCESS AND FINAL PRODUCT

This quantitative, experimental, exploratory and descriptive study was based on data from the research project of the Scientific and Technological Network for Food Bioavailability Studies (REBIAL).

The research was carried out in the Experimental Technology and Nutrition, Sensory Analysis, Chemical Analysis and Bromatology Laboratories of the Life Sciences Department (DCV) of the State University of Bahia (UNEB), campus I, Salvador.

4.1 Raw materials

The samples of mangalô, caupi and andu beans were obtained from family farmers in the city of Cruz das Almas, Bahia, and stored frozen in the Experimental Technology and Nutrition laboratory of the DCV - UNEB. The other ingredients (rice flakes, oat flakes, raisins, crystal sugar, maltodextrin, salt, soya oil, soya lecithin and citric acid) used in the production of the formulations were obtained from local shops in Salvador-BA.

The invert sugar used in the formulation as one of the ingredients of the binding solution was produced in the Experimental Technology and Nutrition laboratory of the DCV at UNEB.

4.1.1 Making invert sugar

Invert sugar was developed in the Experimental Technology and Nutrition laboratory at UNEB's DCV from a mixture of refined sugar, water and citric acid. The mixture was boiled over low heat, reaching a maximum temperature of 114°C for 20 minutes, during which time the mixture was not stirred, as this would increase the risk of crystallization.

After reaching room temperature, the inverted sugar was stored in a glass jar and sealed tightly.

4.1.2 Raw material processing

The formulations were developed through preliminary tests in the Experimental

19

Technology and Nutrition laboratory at UNEB's DCV, using a basic cereal bar formulation. Three cereal bar formulations were prepared with the addition of different amounts of cowpea flour (FFC), mango bean flour (FFM) andandu bean flour (FFA), varying the proportions between the ingredients, in order to study the effects of the presence of these flours on the organoleptic characteristics and nutritional potential of the cereal bars developed.

4.1.3 Manufacturing process and formulations

The following will describe the stages in the process of producing bars based on mangalô, andu and caupi beans with equivalent concentrations. Initially, the processing of the raw material will be discussed, in particular the obtaining of bean flour for incorporation into the bar formulation.

To make the cereal bars (Table 3), the ingredients were divided into two groups: the dry ingredients (oat flakes, rice flakes, bean flour and raisins) and the wet ingredients (soybean oil, crystal sugar, water, maltodextrin, salt, soy lecithin and invert sugar). The cost of producing the cereal bars was taken into account by choosing low-cost ingredients, emphasizing the importance of developing a product that is accessible to family farmers.

The processing steps for bitter mangrove bean (FM), caupi bean (FC) and andu bean (FA) flours were followed according to the flowchart (Figure 7).

Figura 7: Stages in the process of making FM, FC and FA flours.

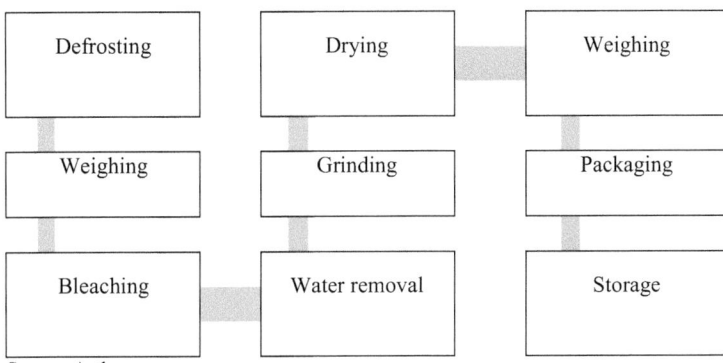

Source: Authors

The frozen mangrove, caupi and andu bean samples underwent a blanching process, consisting of immersing the product for 5 minutes in water at a temperature of 95-100°C (LIMA, et al., 2004), draining off the hot water and then immersing it in cold water. Once the water had been removed, it was ground in a blender (Figure 8) and dried in the oven at a maximum temperature of 160°C (Figure 9), repeating the grinding and drying process until the flour was obtained (Figure 10).

Figura 8: Bitter mangrove beans (*Lablab purpureus* (L.) Sweet) ground in a blender before drying.

Source: Authors

Figura 9: Bitter mangrove beans (*Lablab purpureus* (L.) Sweet) ground in a blender and dried in a domestic oven.

Source: Authors.

Figura 10: Bitter mangrove bean flour (*Lablab purpureus* (L.) Sweet).

Source: Authors

Table 3. Dry ingredients and binding agents used in the basic formulation of cereal bars from FFM, FFC, FFA.

Ingredients	Mass (g)[1]			Percentage
	BFFM	BFFC	BFFA	
Rice flakes	31,28			19,1
Oat flakes	28,15			17,2
Bean Flour	34,4			21
Banana raisins	70			42,7
Total dry ingredients	*163,84*			*100*
Invert liquid sugar	80			55,8
Crystal sugar	15			10,5
Soy lecithin	1,92			1,3

22

Soybean oil	18,24	12,7
Salt	1,37	0,9
Maltodextrin	10,45	7,3
Water	16,5	11,5
Total giants	***143,48***	***100***

BFFM: Bitter mangrove bean flour cereal bar;

BFFCCaupi bean flour cereal bar;

BFFA: Andu bean flour cereal bar.

Source: Authors

The formulations were developed following the steps described in the optimized process flowchart shown in Figure 11:

Figura 11:   Flowchart of the steps for formulating cereal bars from FFM, FFC, FFA.

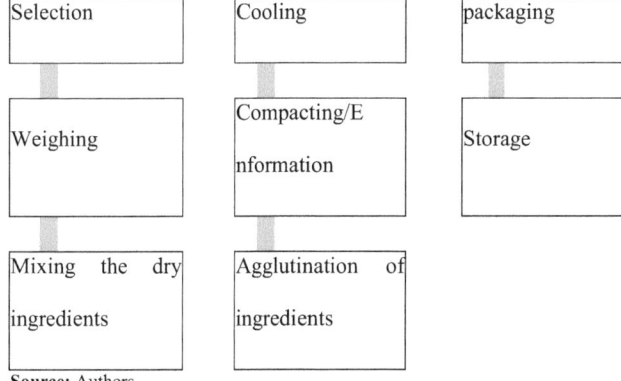

Source: Authors

The stages of the manufacturing process are described in detail below:

1.   Selection: the raw materials were selected for their aroma, color, texture, packaging integrity and expiration date;

2.   Weighing: the raw materials were weighed and portioned;

3.   Mix dry ingredients: oat flakes, rice flakes, mangalô or andu or caupi bean flour, dehydrated banana raisins.

4.   Agglutination of the ingredients: the raw materials for the syrup (inverted liquid

23

sugar, water, crystal sugar, maltodextrin, soybean oil) were mixed and dissolved over the heat until boiling, so that the syrup remained homogeneous, removed from the heat and added soy lecithin, salt and banana raisins, returning to the heat until it reached a maximum temperature of 105°, stirring, removed from the heat and added to the mixture of dry ingredients.

5.   Compacting and shaping: the product was compacted and placed in rectangular molds (Figure 12);

6.   Cooling: after acquiring a characteristic consistency, the mixture was cut into rectangular bars;

7.   Packaging: the cereal bars were packaged in metal-framed film.

8.   Storage: the cereal bars were stored at room temperature and in dry, appropriate places.

Figura 12:   Shaping and compactness of mangrove bean flour cereal bars.

Source: Authors

5 CHARACTERIZATION OF THE FINAL PRODUCTS

5.1 Chemical and sensory analysis methods

Physico-chemical and sensory analyses are carried out to characterize the products. The methods and procedures used to carry out the laboratory tests are described below.

5.1.1 Chemical analysis of the cereal bar

The chemical composition was analyzed according to the methods of the Adolfo Lutz Institute - IAL and AOAC. Moisture analysis was carried out by direct drying in an oven at 105°C, total ash: by incinerating the product at a temperature of 500-550°C in a muffle furnace, lipids by the Soxhlet method, proteins by the Kjeldahl method, fiber by acid detergent, carbohydrates by difference. And determination of the total energy value: according to the ATWATER conversion factors: 4 kcal g^{-1} for proteins, 4 kcal g^{-1} for carbohydrates and 9 kcal g^{-1} for lipids (BRASIL, 2005), of the cereal bar formulations made with FFC, FFM and FFA.

5.1.2 Sensory analysis of cereal bars

The sensory analysis of cereal bars made from mangalô, caupi and andu bean flour was carried out in the Sensory Analysis Laboratory at the DCV-UNEB. 60 consumers took part, of both sexes, with or without an employment contract with the institution. In order to carry out the sensory tests, this study was approved by the Ethics Committee No. 1.145.758. Each taster was given the "Informed Consent Form" (Appendix A) and the product evaluation form (Appendix B), in which the purpose of the analysis was presented and consent to take part was requested. The form was presented in two copies, one for the taster and one for the research control.

The analysis was carried out in an appropriate place, under natural light, with 03 samples (one from each treatment under study) and presented to the tasters on disposable plates, duly coded with three-digit numbers chosen at random. The sample was also offered mineral water at room temperature.

Acceptability was assessed in terms of the following attributes: appearance, overall quality, aroma, flavor and texture, using a nine-point verbally structured hedonic scale. The same form included a scale to assess the consumer's attitude in a hypothetical purchase situation.

5.1.3Statistical Analysis

The data obtained from the chemical determinations and the sensory evaluation were submitted to analysis of variance (ANOVA), adopting a significance level of 5% ($P \leq 0.05$) and contrast between the means by the Tukey test, using SAS *software* version 9.1.

5.2 Characterization of the final products

5.2.1Centesimal composition

The results regarding the centesimal composition of the cereal bars analyzed are shown in Table 4.

Table 4: Results of the centesimal composition of cereal bars from FFC, FFA, FFM.

PARAMETERS	FORMULATIONS		
	BFFC	BFFA	BFFM
HUMIDITY	12,79±0,21[a]	12,83±0,66[a]	12,77±0,21[a]
GREY	2,16±0,36[a]	2,45±0,23[a]	2,11±0,31[a]
LIPIDS	8,64±0,87[a]	8,58±1,63[a]	11,61±3,41[a]
PROTEINS	5,99±0,47[a]	6,18±0,42[a]	6,34±0,16[a]
FIBERS	2,93±1,61[a]	2,93±0,06[a]	1,10±0,13[a]
CARBOHYDRATES[4]	67,30±1,62[a]	67,22±1,99[a]	66,07±3,46[a]

[1] Values are mean ± standard deviation.

[2] Averages followed by the same letters in the columns do not differ by the Tukey test at a significance level of 5% ($p \leq 0.05$).

[3] BFFC = caupi bean flour cereal bar, BFFA = andu bean flour cereal bar, BFFM = mangrove bean flour cereal bar. 4Carbohydrate content obtained by difference.

Source: Authors

Based on this data, we can say that the FFC, FFA and FFM cereal bars did not show any significant differences ($p < 0.05$) in relation to the parameters observed.

Table 4 shows that the moisture content of the formulations is less than 15%, the

26

limit established by CNNPA Resolution No. 12 of 1978 for products based on cereals and derivatives, allowing the products to have a longer shelf life, guaranteeing the texture, chemical and microbiological stability of the cereal bars (LEAL et al., 2013). However, according to Cecchi (2007) the moisture content of cereals should be below 10%. In the study by Sousa et al. (2012), the moisture content found was 10.88% and 11.28% in cereal bar formulations based on FFC.

The moisture contents found in the FFC analyzed in other studies were (11.61 and 11.85 g $100g^{-1}$), results close to those found in a study by Santos et al. (2009) with common bean flour (11.7 g $100g^{-1}$) and lower than those obtained in the analysis of raw common bean flour (17.60 g $100g^{-1}$) and cowpea flour (14.3-15.8 g $100g^{-1}$) (GOMES et al., 2006; GOMES et al., 2012; LEAL et al., 2013).

The total ash values in cereals, which are related to the mineral content of the food, are in line with Cecchi (2007) and can vary from 0.3 to 3.3%, indicating that the cereal bars developed here have a good presence of minerals.

According to Table 4, the values found for lipids are higher than those recommended by Cecchi (2007), ranging from 3% to 5%, and similar to those found by Sousa et al. (2012). However, the values found are within the range of lipid content of conventional products found on the market, from 4.0 to 12.0% (FREITAS, MORETTI, 2006).

All the bars showed a lower concentration of crude fiber compared to the values published in some of the literature consulted. This variation can be attributed to losses due to hydrolysis in the method used, and due to the insoluble fraction of fiber in green beans, which represents around 75%, with less mature products having a lower amount of fiber (SALGADO et al., 2005). Crude fiber has no nutritional value, but provides the necessary tool for peristaltic bowel movements, with the content of crude fiber in cereals and cereal products ranging from 0.00 to 2.2% (CECCHI, 2007).

The FFC cereal bar sample had a lower protein content compared to the cereal bars developed in the study by Sousa et al. (2012), with formulations containing 5.25%

and 7.5% FFC. However, in the study by Sousa et al. (2012), the increase in protein concentration can be explained by the fact that the dry ingredients contained cornflour cookies, which may contain milk.

The carbohydrate contents were obtained by difference, 67.30%, 67.22%, 66.07%, for BFFC, BFFA, BBFM, respectively. They were similar to those found in the literature.

Table 5 shows the Total Energy Value (TEV) of the cereal bars developed.

Table 5: Total energy value (kcal 100g^{-1}) of FFC, FFA and FFM cereal bars.

TOTAL CALORIFIC VALUE	
BFFC	382.64 kcal
BFFA	382.54 kcal
BFFM	398.53 kcal

Source: Authors

The FFC, FFA and FFM cereal bar formulations developed had a moderate energy value (1.5 to 4 kcal g^{-1}), according to the *Centers for Disease Control and Prevention* classification *(2005),* attributed mainly to the amount of lipids and carbohydrates in the samples, due to the high percentage of raisins and sugar used as a binding agent. These cereal bars can be alternatives for those who need a high-calorie diet.

5.2.2Sensory Analysis

The averages obtained from the sensory analysis, submitted to analysis of variance, for each of the formulations and for the five attributes analyzed (appearance, aroma, taste, texture and overall quality) are shown in Table 6.

Table 6: Average scores obtained for the sensory attributes and their respective standard deviations of the cereal bars from the FFC, FFA, FFM.

FORMULATIONS[3]	APPEARANCE	AROMA	TASTE	TEXTURE	QUALITY GLOBAL
FFC	6,50±1,59[a]	5,42±1,57[a]	6,15±1,58[a]	6,52±1,08[a]	**6,23±1,10[b]**

28

FFA	$6,38\pm1,32^a$	$5,58\pm1,38^a$	$\mathbf{5,58\pm1,42^b}$	$6,48\pm1,16^a$	$\mathbf{5,83\pm1,05^{b,a}}$
FFM	$6,37\pm1,55^a$	$5,42\pm1,47^a$	$6.13\pm1,93^a$	$6,53\pm1,25^a$	$6,35\pm1,25^a$

[1] Values are: mean ± standard deviation.

[2] Averages followed by equal letters in the columns do not differ by the Tukey test, at a significance level of 5% ($p\leq0.05$).

[3] FFC = caupi bean flour cereal bar, FFA = andu bean flour cereal bar, FFM = mangalô bean flour cereal bar.

Source: Authors

The cereal bar formulations with FFC, FFA and FFM generally showed good sensory acceptance. The averages ranged from "indifferent" to "moderately liked" on the 9-point structural hedonic scale, showing that the products obtained similar results in all the sensory characteristics evaluated.

Based on the data presented in Table 6, the samples did not differ significantly ($p<0.05$) in terms of appearance, aroma and texture, i.e. it can be said that the FFC, FFA and FFM cereal bar formulations were homogeneous in terms of these attributes, with good acceptance by the judges. However, there were significant differences in the taste and overall quality attributes ($p>0.05$).

The formulation that showed a sensory alteration in terms of taste was the FFA cereal bar, with the lowest acceptance (5.58%), given that andu beans are appreciated because they are more palatable. A possible explanation for this result could be the ineffectiveness of the heat treatment, which inactivates enzymes and improves the taste and aroma.In terms of overall quality, the FFM cereal bar showed better quality than the FFC cereal bar, but the FFA cereal bar did not differ from the FFC and FFM cereal bars.

Of all the averages, the aroma characteristic received the lowest scores, ranging from "indifferent" to "I liked it slightly". This result shows that the product could be improved. The addition of natural flavoring raw materials can improve the product's aroma, as this process tends to add fruits, fruit juices, nuts or spices that make its aroma more pleasant.

To assess purchase intention (Figures 13, 14, 15), a five-point scale was used. The results show that the majority of consumers would certainly buy cereal bars. Figure

16 shows the general results obtained in the sensory analysis of the three samples for the purchase intention test, in which consumers expressed how often they would consume each of the samples.

Figure 13: Results of the affective test by attitude scale or intention to buy the FFC cereal bar.

Figure 14: Affective test results by attitude scale or intention to buy the FFA cereal bar.

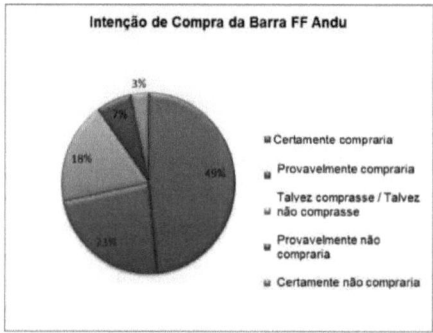

Source: Authors

Figura 15: Results of the affective test using an attitude or purchase intention scale for the FFM cereal bar.

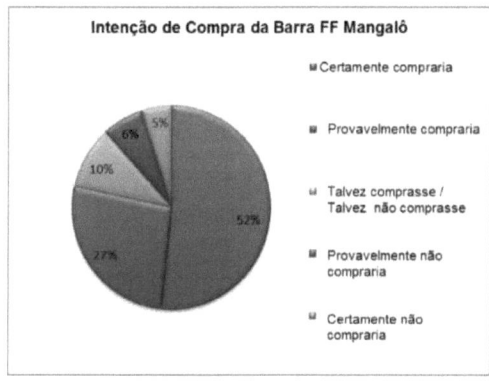

Source: Authors

Figura 16: General results of the affective test using an attitude scale or intention to buy cereal bars based on the FFC, FFA, FFM.

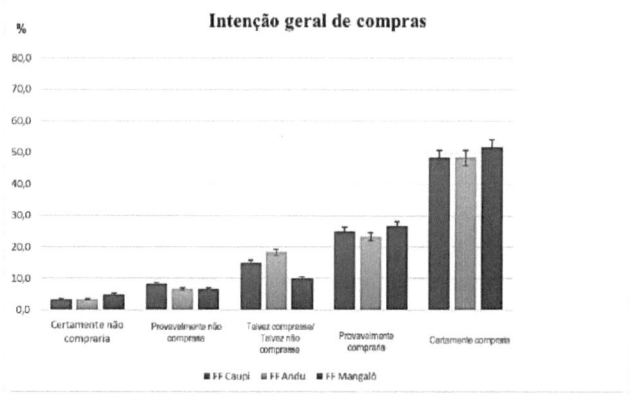

It was observed that 52% of consumers said that they would certainly buy the FFM cereal bar, compared to the FFC and FFA cereal bars, which received 49% of the affirmation. However, 5% of consumers said they would certainly not buy FFM cereal bars, while 3% said they would not buy FFC and FFA cereal bars.

31

6. FINAL CONSIDERATIONS

The results obtained from the analysis of the cereal bars were considered satisfactory, allowing us to conclude that the flours of caupi, andu and mangalô beans can be used as an ingredient in the preparation of cereal bars, with the FFM cereal bar obtaining the best acceptance. The centesimal analysis of the three cereal bars produced showed their richness in nutrients and calories.

The technique used to obtain the flours is easy to apply, but studies are needed to determine the losses in nutritional value when the product is processed.

The products developed value local culture and food habits, adding a regional raw material to a food that is widely consumed in the health food market.

REFERENCES

AHMAD, Mushtaq et al. A review on Oat (*Avena sativa* L.) as a dual-purpose crop. **Scientific Research And Essays,** Nigeria, v. 9, n. 4, p.52-59, feb. 2014. ISSN: 1992-2248.

ALMEIDA, Ana Claudia Santana de. **Study of the continuous process for producing invert sugar enzymatically.** 2003. 99 f. Dissertation (Master's Degree) - Postgraduate Course in Chemical and Biochemical Processes, Chemical Engineering, Federal University of Pernambuco, Recife, 2003.

AZEVEDO, Ruberval Leone; RIBEIRO, Genésio Tâmara; AZEVEDO, Clàudio Luiz Leone. Guandu Beans: A Multipurpose Plant. **Revista da Fapese,** v. 3, n. 2, p.81-86, jul./dez. 2007.

BRASIL. Eduardo Alves Melo (Ed.). **Alimentos Regionais Brasileiros.** 2. ed. Brasilia: Ministério da Saùde, 2015. 484 p. ISBN:978-85-334-2145-5.

BRAZIL. Ministry of Agriculture, Livestock and Supply. Normative Instruction n° 8, of June 2, 2005. Technical regulation on the identity and quality of wheat flour. **Official Gazette of the Federative Republic of Brazil**, Brasilia, DF, n. 105, p. 91, June 3, 2005. Section 1.

BRAZIL. Resolution No. 263, of July 22, 2005. **Resoluçâo RDC N° 263, de 22 de Setembro de 2005**: Regulamento tènico para produtos de cereais, amidos, farinhas e farelos. Brazil: D.O.U - Diàrio Oficial da Uniao; Poder Executivo, July 22, 2005. Available at: <http://portal.anvisa.gov.br/wps/wcm/connect/1ae52c0047457a718702d73fbc4c6735/RDC_263_2005.pdf?MOD=AJPERES>. Accessed on: Feb. 18, 2016.

BRAZIL. Law No. 11.326, of July 24, 2006. Establishes the guidelines for the formulation of the National Policy for Family Farming and Rural Family Enterprises. **Official Gazette of the Federative Republic of Brazil**, July 25, 2006. Available at:<http://www.planalto.gov.br/ccivil_03/_ato2004-   2006/2006/lei/l11326.htm>. Accessed on: May 15, 2016.

BRAZIL. Ministry of Health Food guide for the Brazilian population. Brasilia: Ministry of Health; 2. ed. 175 p., 2014

BRIGID MCKEVITH (United Kingdom). British Nutrition Foundation. Nutritional aspects of cereals. **Nutrition Bulletin,** London, n. 29, p.111-142, jun. 2004.

CHAVES, Michela Okada; BASSINELLO, Priscila Zaczuk. **Beans in human nutrition**. 2014. Available at:

<http://ainfo.cnptia. embrapa.br/digital/bitstream/item/ 123450/1/p15.pdf>. Accessed on: June 6, 2016.

CECCHI, H. M. **Theoretical and Practical Foundations of Food Analysis**.

Campinas: Editora Unicamp. 2^a ed. 2007

COUTINHO, Ana Paula Cerino. **Production and characterization of maltodextrins from cassava and sweet potato starches.** 2007. 151 f. Thesis (Doctorate) - Agronomy Course, Universidade Estadual Paulista "Julio de Mesquita Filho", Botucatu, 2007.

DEGÀSPARI, Clàudia Helena; BLINDER, Elsa Wasserman; MOTTIN, Fatima. Nutritional profile of cereal bar consumers. **Visâo Acadêmica,** Curitiba, v. 9, n. 1, p.49-61, mar. 2008. ISSN 1518-5192.

UNITED STATES. Food and Drug Administration. Department Of Health And Human Services. **Maltodextrin.** 2015. Available at:

https://www.accessdata.fda.gov/scripts/cdrh/cfdocs/cfcfr/cfrsearch.cfm?fr=184.1444 > . Accessed on: March 30, 2016.

FERREIRA, Carlos Magri; PELOSO, Maria José del; FARIA, Luis Clàudio de.

Cultivation of the Common Bean: Market and commercialization. **Embrapa Arroz e Feijâo,** v. 2, jan. 2003.

FOOD INGREDIENTS BRAZIL. **Emulsifiers**. 2013. Available at: <http://www.revista-fi.com/materias/324.pdf>. Accessed on: May 29, 2016.

FRAGON. **Soya lecithin powder.** Technical material produced by Fragon. Available at: <http://cdn.fagron.com.br/doc_prod/docs_10/doc_929.pdf>. Accessed on: March 22, 2016.

FREIRE FILHO, Francisco Rodrigues; ROCHA, Maurisrael de Moura. **Green grains.** Prepared by the Embrapa Agency for Technological Information. Available at: <http : //www.agencia. cnptia.embrapa. br/gestor/feij ao-caupi/arvore/CONTAG01 _ 76_510200683537.html>. Accessed on: March 30, 2016.

FREITAS, Antônio Carlos Reis de. **The economic importance of cowpeas.** 2011. Prepared by: Embrapa Agency for Technological Information.

Available at: <http://www.agencia.cnptia.embrapa.br/gestor/feij ao - caupi/arvore/CONTAG01_14_ 510200683536.html>. Accessed on: March 23, 2016.

FROTA, Karoline de Macêdo Gonçalves et al. Utilization of cowpea flour (*Vigna unguiculata* L. Walp) in the preparation of bakery products. **Food Science and Technology (campinas),** Campinas, v. 30, p.44-50, May 2010.

GALLI, D. C.; BILHALVA, A. B.; RODRIGUES, R. S.; RODRIGUES, L. S. Influence of syrup composition on the physico-chemical characteristics of raisin peaches. **Revista Brasileira de Agrociência**, v. 2, n. 3, p. 179-182, Sept./Dec. 1996.

GARDEN-ROBINSON, J.; MCNEAL K.All About Beans. NDSU -North Dakota State University, 16 p., 2013.

GÓES, A. C. P., CAVALCANTEE. S. The cowpea in numbers. Embrapa Amapa. 2013. Available at: <https://www.infoteca.cnptia.embrapa.br/ bitstream/doc /975559/1/CPAFAP2013FolderOFEIJAOCAUPIEMN UMEROSPA

RAPUBLICATION.pdf>Accessed: 19 Mar. 2016

GOWDAK, M. M. G. Sodium content in food < http://www.sbh.org.br/geral/ actualidades-teor- de-sodio-na-alimentacao.asp> Accessed on: 11 Mar 2018

GUILHOTO,Uoaquim et al. **A Importância da Agricultura Familiar no Brasil e em seus Estados (Family Agriculture's GDP in Brazil and in It's States**), 2007. V National Meeting of the Brazilian Association of Regional and Urban Studies, 2007

GUIMARÂES, M. M.; SILVA, M. S. Nutritional quality and acceptability of cereal bars with added murici-passa fruit. **Revista do Instituto Adolfo Lutz**, Sao Paulo, v.68, n.3, p.426-433, 2009.

GUTKOSKI, L.C. et al. Development of an oat-based cereal bar with high dietary fiber content. **Ciência e Tecnologia Alimentos**, Campinas, v.27, n.2, p. 355-363, 2007.

GUTKOSKI, L.C.; TROMBETTA, C. Evaluation of dietary fiber and glycan content in oat cultivars. **Ciência e Tecnologia de Alimentos**, v. 19, n. 3, p. 387-390, 1999.

HAMMOND, Earl G. et al. Soybean Oil. In: SHAHIDI, Fereidoon. **Bailey's Industrial Oil and Fat Products:** Edible Oil and Fat Products: Chemis. 6. ed. New Jersey: John Wiley & Sons, 2005. Chap. 13. p. 577-642.

Agricultural Research Company of Minas Gerais. **Unconventional vegetables**: an alternative for diversifying food and income for family farmers in Minas Gerais. Minas Gerais: Publications Department, 20015. 24 p.

IBGE. Household Budget Survey 2008-2009: **Consumption Analysis Personal Nutrition in Brazil.** Rio de Janeiro: Brazilian Institute of Geography and Statistics, 2011

ADOLFO LUTZ INSTITUTE. **Normas Analiticas do Instituto Adolfo Lutz**: Métodos quimicos e fisicos para Anâlise de alimentos. 3. ed. Sao Paulo: 2005

LIMA, Eliza Dorotea P. de A. et al. (Org.). **Green cowpea (*Vigna unguiculata (*L.) Walp.):** Post-harvest aspects, minimal processing, canned processing. Joao Pessoa: University, 2004.

LOUIZE, Jaqueline. **Cereal bars, protein bars, energy bars, diet bars, light bars, ... get to know the differences, the pitfalls, the 100% vegetable (vegan) ones and which ones are worth eating**. Available at: <http://ecocheervegan.com/nutricao-vegetariana/191- conhecera-as-barras-de-cereais>. Accessed on: 05 Apr. 2016

MARQUES, Tamara Rezende. **Technological use of acerola waste:** flours and cereal bars. 2013. 103 f. Dissertation (Master's Degree) - Agrochemistry Course, Federal University of Lavras, Lavras - Mg, 2013.

MARQUEZI, Milene. **Physico-chemical characteristics and evaluation of technological properties of common bean (*Phaseolus vulgaris L.*).** 2013. 115 f. Dissertation (Master's Degree) - Postgraduate Course in Food Science, Center for Agricultural Sciences, Federal University of Santa Catarina, Florianólolis, 2013.

MIZUBUTI, I.Y. et al. Evaluation of the use of ground raw guandu beans (Cajanus cajan (L) Millsp) on the indirect productivity indices of broiler chickens. **Semina Ciências Agràrias**, Londrina, v. 16, n. 1, p. 56-63, 1995.

NATABIRWA, H.N.; KATENDE D.; LUNG'AHO M.. **Bean recipes:** best food choice for the adventurous cook. Uganda: National Agricultural Research Laboratories (NARL/NARO), International Center For Tropical Agriculture (CIAT), Pan-Africa Bean Research Alliance (PABRA), 2014. 44 p. Available at: <https://cgspace.cgiar.org/handle/10568/71054>. Accessed on: March 18, 2016.

NEPA - NÙCLEO DE ESTUDOS E PESQUISAS EM ALIMENTAÇÂO. Brazilian Table of Food Composition (TACO). 4ª edrev. e ampl. Campinas: NEPA - UNICAMP, 2011. 161 p.

OETTERER, Marilia; SARMENTO, Silene Bruder Silveira. Properties of sugars. In: OETTERER, Marilia; REGITANO-D'ARCE, Marisa Aparecida Bismara; SPOTO, Marta Helena Fillet. **Fundamentals of Food Science and Technology.** Barueri: Manole, 2006. Chap. 4. p. 135-192.

PHILIPPI, Sônia Tucunduva. Sugars. In: PHILIPPI, Sônia Tucunduva. **Nutrition and Dietetic Techniques.** 3. ed. Barueri: Manole, 2014. Chap. 14. p. 185-198.

PIOVESANA, Alessandra. **Preparation and acceptability of cereal bars with grape pomace.** 2011. 59 f. TCC (Graduation) - Higher Education Course of Food Technology, Federal Institute of Education, Science and Technology of Rio Grande do Sul, Bento Gonçalves, 2011.

PODADERA, Priscilla. **Study of the properties of inverted liquid sugar processed with gamma radiation and electron beam.** 2007. 108 f. Thesis (Doctorate) - Nuclear Technology - Applications Course, Nuclear and Energy Research Institute, Sao Paulo, 2007.

PORTAL BRASIL. Family farming produces 70% of the food consumed by Brazilians. jul. 2015. Available at: <http://www.brasil.gov.br/economia-e-emprego/2015/07/agricultura-familiar-produz-70-dos-al alimentos-consumidos-por-brasileiro>. Accessed on: March 20, 2016.

RIBEIRO, Valdenir Queiroz (Ed.). In: Embrapa Meio-Norte. Cultivation of cowpea (*Vigna unguiculata* (L.) Walp). **Sistemas de Produçâo.**Teresina, v.2, p.1-110, dec. 2002. ISSN 1678-0256.

RUAS, Joao Figueiredo. Beans. In: Companhia Nacional de Abastecimento **Perspectivas Para A Agropecuâria: Safra 2015/2016,** Brasilia, v. 3, p.43-49, jul.

2015. Annual. ISSN 2318-3241. Available at:

<http : //www.conab .gov.br/OlalaCMS/uploads/arquivo s/15_09_24_11 _44_5 0_perspe ctivas_agropecuaria_2015-16_-_produtos_verao.pdf>. Accessed on: 18 Mar. 2016.

RUBATZKY, V.E.; YAMAGUCHI, M. **World vegetables**: principles, production and nutritive values. 2nd Edition. Chapman & Hall, New York, United States. 1997, 843 pp.

SALVADOR, Carlos Alberto. **Beans:** Analysis of the Agricultural Situation. 2015. In: State Secretariat for Agriculture and Supply. Available at: <http://www.agricultura.pr.gov.br/arquivos/File/deral/Prognosticos/2016/_feijao_201 5_16.pdf>. Accessed on: March 18, 2016.

SAMPAIO, Camila Ramos Pinto. **Development and study of the sensory and nutritional characteristics of iron-fortified cereal bars.** 2009. 88 f. Dissertation (Master's Degree) - Postgraduate Course in Food Technology, Federal University of Paranà, Curitiba, 2009.

SANTOS, Juliana Ferreira dos. **Evaluation of the nutritional properties of cereal bars made with green banana flour**. 2010. 70f f. Dissertation (Master's Degree) - Food Science Course, University of Sao Paulo, Sao Paulo, 2010.

SILVA, Barbara Cristina Dantas da; COSTA, Ana Elisa Del'arco Vinhas. Socio-productive diagnosis of family farmers cooperating with the family farmer's cooperative in the recôncavo territory of Bahia - Brazil.

COOAFATRE. **Magistra,** Cruz das Almas, v. 24, n. 2, p.151-159, abr/jun. 2012. ISSN 2236-4420.

SOUSA, Luska Grazielle Macêdo de et al. **Preparation of a cereal bar based on cowpea flour (*vigna unguiculata* l. walp).** Available at: <http://leg.ufpi.br/21 sic/Documentos/RESUMOS/Modalidade/PIBITI/Iuska Grazielle.pdf>. Accessed on: February 14, 2016.

SOUSA, Viviane. **Cereal bars gain strength**. Available at: <http://www.sm.com.br/detalhe/barras-cereais-ganham-forca>. Accessed on: 14 Feb. 2016.

TRAMUJAS, Janaina Melati. **Use of different binding agents in the development of salted cereal bars with chia (*Salvia hispânica l.*).** 2015. 125 f. Dissertation (Master's Degree) - Food Technology Course, Universidade Tecnològica Federal do Paranà Londrina, 2015.

APPENDIX A - Informed Consent Form

BAHIA STATE UNIVERSITY- UNEB

DEPARTMENT OF LIFE SCIENCES - CAMPUS I

COLLEGIATE OF NUTRITION

INFORMED CONSENT FORM

THIS RESEARCH WILL FOLLOW THE CRITERIA FOR ETHICS IN RESEARCH WITH HUMAN BEINGS IN ACCORDANCE WITH RESOLUTION NO 466/12.

OF THE NATIONAL HEALTH COUNCIL

we invite you to take part in the research project "DEVELOPMENT OF CEREAL BARS BASED ON LEGGEMINES FROM FAMILY FARMERS IN THE CITY OF CRUZ DAS ALMAS-BAHIA", whose general objective is to develop cereal bars from legume flours in order to help the social, economic and sustainable development of family farmers in the city of Cruz das Almas-Bahia.

This is a Course Conclusion Work, developed by the student Camila de Oliveira Barros and supervised by Prof.[a] Dr.[a] Katia Elizabeth de Souza Miranda, of the Bachelor's Degree in Nutrition of the Life Sciences Department of the State University of Bahia.

Their participation is voluntary and will take the form of a sensory analysis of the cereal bars developed and the completion of a questionnaire on taste, texture, appearance and their intention to buy the product studied. The samples will be ready for consumption and the results of the survey will show the degree of acceptability of the cereal bars. Participation does not involve any financial costs or rewards for the participants. You may withdraw your consent at any time. Your refusal will not affect your relationship with the researcher or the institution. The results of the research will be analyzed and published, but your identity will not be disclosed and will be kept confidential.

CONFIDENTIALITY OF THE RESEARCH: Participants must guarantee the confidentiality of the data involved in the research.

If you agree with the above, please sign this "Informed Consent Form" in the place indicated below. Thank you in *advance for* your cooperation.

Me   ,

i inform you that i have been duly informed about what the researcher wants to do and why he/she needs my cooperation, and i have understood the explanation. I therefore agree to take part in the

40

research of my own free will, knowing that the research is confidential, that I will not earn anything and that I can leave whenever I want. I consent to the results obtained being presented and published in scientific events and articles, provided that I am not identified. This document is issued in two copies, both of which will be signed by me and the researcher, with one copy remaining with each of us.

Salvador, _____ 2016

Volunteer

Camila de Oliveira Barros Researcher Bachelor of Nutrition UNEB/DCV

Prof. Dr. Katia Elizabeth de Souza Miranda

Advisor - UNEB/DCV

APPENDIX B - Acceptance and purchase intention test sheet and acceptability assessment of cereal bars made from FFM, FFC, FFA

ACCEPTANCE TEST

N ome : Sex : _____ Age : _____

Please taste the samples coded from left to right and using the scale below, describe how much you liked or disliked each sample, according to the attributes listed.

(9) = I really liked it

(8) = I liked it a lot

(7) = I liked it moderately

(6)= I really liked it

(5) = indifferent

(4) = I strongly dislike

(3)=moderately disliked

(2) = I really disliked it

(1) = extremely dislike

Comments:

Samples	Attributes				
_	Appearance	Aroma	Flavor	Texture	Overall quality
_	Appearance	Aroma	Flavor	Texture	Overall quality
	Appearance	Aroma	Flavor	Texture	Overall quality
_					

Purchase intent:

N^0 Sample

I would certainly buy it   _____ _____

I'd probably buy it   _____ _____

Maybe I would / Maybe I wouldn't   _____ _____

I probably wouldn't buy it   _____ _____

I certainly wouldn't buy it   _____ _____

ACCEPTANCE TEST

Name: Sex: Age:

Please taste the coded samples from left to right and using the scale below, describe how much you liked or disliked each sample, according to the attributes listed.

(9) = I really liked it

(8) = I liked it a lot

(7) = I liked it moderately

(6)= I really liked it

(5) = indifferent

(4) = I strongly dislike

(3)=moderately disliked

(2) = I really disliked it

Samples	Attributes				
_	Appearance	Aroma	Flavor	Texture	Overall quality
_	Appearance	Aroma	Flavor	Textuia	Overall quality
	Appearance	Aroma	Flavor	Texture	Overall quality
_					

(1) = extremely dislike						
Comments:_____	№ Sample					

Purchase intent:

I would certainly buy it

I'd probably buy it

Maybe I would / Maybe I wouldn't

I probably wouldn't buy it

I certainly wouldn't buy it

authors' names

CAMILA DE OLIVEIRA BARROS

http://lattes.cnpq.br/2517848703479650

Master's student in Nutrition and Food Science with an emphasis on Analysis of health risks associated with food at the Université Paris-Saclay (Agroparistech), France (2018). Voluntary Scientific Initiation in the Food Bioavailability research project - REBIAL at the State University of Bahia (2016).Graduated in the area of Nutrition at the State University of Bahia (2016). Participation in the Science without Borders Program as a CAPES scholarship holder (2014).

KATIA ELIZABETH DE SOUSA MIRANDA

http://lattes.cnpq.br/5053104510054359

She holds a PhD in Food Science and Technology from the Federal University of Paraiba (2011), an undergraduate degree in FOOD ENGINEERING from the Federal University of Paraiba (1986) and a master's degree in Food Science and Technology from the Federal University of Paraiba (1991). She is currently Full Professor of Basic, Technical and Technological Education at the Federal Institute of Bahia and Adjunct Professor at the State University of Bahia. She has experience in the area of Nutrition, with an emphasis on Vegetable Product Technology, working mainly on the following subjects: vegetable processing and experimental technology in Nutrition.

WAGNA PILER CARVALHO DOS SANTOS

http://lattes.cnpq.br/7745470765033035

She holds a PhD in Chemistry from the Federal University of Bahia - UFBA (2007), a master's degree in Chemistry from UFBA (2003), a degree in Chemistry from UFBA (2001) and a degree in Food Technology from the Federal Technical School of Chemistry of Rio de Janeiro, now IFRJ. She taught the Food Technician course at the Federal Center for Technological Education of Paranà-CEFET/PR, now UTFPR. She is currently a professor at the Federal Institute of Education, Science and Technology of Bahia (IFBA). She has experience in the field of Chemistry, with an

43

emphasis on Analytical Chemistry, working mainly on the following subjects: spectroanalytical techniques, ICP OES, sample preparation, food,

legumes and essential and toxic elements. She has been the National Coordinator of the Intellectual Property (IP) Concepts and Applications course at PROFNIT since its inception.

LIGIA REGINA RADOMILLE DE SANTANA

http://lattes.cnpq.br/7289150597211694

He has a degree in Food Engineering from the State University of Campinas (1980), a master's degree in Agricultural Sciences from the Federal University of Bahia (2000) and a doctorate in Agricultural Engineering from the State University of Campinas (2009). He is currently a full professor at the State University of Bahia. He has experience in the area of Food Science and Technology, with an emphasis on Technology of Products of Plant Origin, working mainly on the following subjects: Sensory Analysis of Foods, Food Processing and Stability, Physical, Chemical and Chemical Evaluation of Foods, Post-Harvest of Horticultural Products.

Printed by Books on Demand GmbH, Norderstedt / Germany